AF849229

Table of Contents

Chapter 1: Early Measuring Devices

Ancient civilizations, lacking the precision instruments we take for granted, relied on ingenious methods for measurement. Consider the imple balance scale, a testament to early human ingenuity. Variaions existed across cultures, from the elegant Egyptian scales with inricately carved counterweights to the more rudimentary designs used in Mesopotamia. These weren't just for weighing goods in marketplaces; they played a crucial role in medicine, allowing physicians to accurately measure doses of herbal remedies and other substances. The development of calibrated weights, often made from stone or metal, marked a significant step forward, leading to increasingly accurate measurements. The precision of these ancient scales, though limited by today's standards, was remarkable given the available technology, howcasing the dedication to accuracy in various aspects of ancient ife. These scales weren't merely tools; they were symbols of fairness and trade, essential for building societies.

Early rulers, while seemingly simple, represent a fundamental advancement in human civilization. Imagine the effort required to construct a pyramid or a temple without a reliable means of measuring lengths. The earliest examples were often crude markings on bone or wood, but gradually evolved into more sophisticated designs. Egyptian cubit rods, for instance, were standardized lengths, facilitating large-scale construction projects. Roman surveyors used highly refined instruments including grooved rulers, ensuring uniformity across their vast empire. These weren't just for building; they were also crucial for land surveying, mapping territories, and establishing property boundaries – all essential for a functioning state. The precision of these early rulers, while less accurate than modern ones, enabled significant feats of engineering.

The impact of these early measuring devices extends far beyond their immediate applications. The development of accurate scales and rulers spurred innovation in other fields. Accurate measurements were essential for advancements in astronomy, allowing for more precise astronomical observations and calculations. In construction, the ability to measure accurately led to sturdier buildings and more complex structures. In trade, standardized weights and measures fostered fair exchange and economic stability. The early rulers and scales weren't simply tools; they were fundamental building blocks of progress, showcas ing a remarkable ability to adapt and create despite limited technology. The influence these tools had on the development of later precision instruments is undeniable.

Even the seemingly minor improvements—the smoother surface of a ruler allowing for more precise markings, the use of different materials for weights to ensure greater longevity—show a continuous drive for refinement. This iterative process, the constant refinement and adaptation of simple tools, is a testament to human creativity and resourcefulness. The accuracy of these early devices wasn't absolute, but the constant drive for increased accuracy lays the groundwork for the advanced measuring instruments of later eras. Their significance transcends their immediate practical use; they are tangible evidence of humanity's enduring quest for precision and understanding.

The study of these ancient measuring devices offers a valuable insight into the ingenuity and dedication of past societies. These artifacts, often overlooked, are windows into a time when resourcefulness and clever design played a crucial role in overcoming technological limitations. The development of reliable measurement tools wasn't a simple progression; it involved numerous iterations, cultural exchanges, and ingenious problem-solving. They tell a story of human innovation and the slow, painstaking process of building a more accurate and technologically advanced world. By studying these tools, we gain a deeper appreciation of the foundations upon which modern technology is built.

It is in understanding these simpler implements that we truly grasp the enormity of advancements.

1.1 Ancient Scales

Ancient scales, far from being mere weighing instruments, represent a fascinating intersection of engineering, societal structures, and cultural practices. Their evolution mirrors the development of human civilization itself, progressing from simple balancing beams to complex systems reflecting advancements in metallurgy and mathematics. Consider the Egyptian steelyard balance, a precursor to modern scales, where a fulcrum supported a beam with unequal arms. The heavier object was placed on the shorter arm, while known weights were moved along the longer arm until balance was achieved; a testament to the understanding of leverage principles centuries ago. Precision wasn't always paramount; the focus often leaned towards fulfilling basic needs, like fair trade.

These early scales, crafted from readily available materials like wood and stone, often incorporated ingenious design features reflecting the available resources and prevalent construction techniques. In some cultures, intricately carved wooden scales served not only practical purposes but also held symbolic meaning within religious ceremonies or societal hierarchies. The accuracy of these devices varied considerably; a factor influenced by the quality of materials, the skill of the craftsman, and the overall design. Yet, their very existence provided a framework for standardized measurement, however rudimentary, influencing trade, agriculture, and taxation.

The Roman Empire, known for its advanced infrastructure, employed sophisticated steelyard balances and even developed more complex torsion balances. These innovations, often made of bronze or other metals, allowed for more precise measurements and increased capacity. Detailed accounts of Roman engineering projects, like aqueducts and roads, highlight the critical role of accurate weight measurement in procuring and transporting building materials. The consistent demands of a large empire fostered further refinement in scale design, highlighting a clear link between technological progress and governance.

Beyond the Roman world, cultures across the globe developed their own unique weighing systems, showcasing amazing diversity and ingenuity in design. From the delicate balances used in ancient China for weighing precious metals to the robust scales used in the Middle East for trading spices and textiles, variations abound. These systems weren't isolated; trade routes facilitated the exchange of ideas and technology, leading to the gradual adoption and adaptation of different weighing techniques across vast distances. The diverse forms of scales reflect the wide range of materials available and the unique needs of individual societies.

A crucial aspect to note is the connection between scale development and the wider societal context. The use of scales extended far beyond simple commerce; they were deeply embedded in legal systems, often playing a crucial role in administering justice. Fair trade and accurate weights were vital to maintaining social order; scales were not just tools, but integral components of a functioning society. The development of increasingly accurate measuring instruments became inextricably linked with the pursuit of economic fairness and stability. This is clearly illustrated through the rigorous standards implemented by various empires and kingdoms, often with penalties for those who were found using tampered scales or weights.

Examining ancient scales reveals more than just the history of measurement. It provides insight into the ingenuity of our ancestors, their understanding of fundamental scientific principles, and the vital role of precise measurements in shaping the development of human societies. Studying their construction, materials, and the cultural contexts in which they were used offers a valuable perspective on the advancement of both technology and civilization. The seemingly simple act of weighing, made possible by these ingenious inventions, helped to form the economic, social, and even legal foundations of many ancient societies. Their lasting legacy underscores the importance of careful measurement in a world striving for order and efficiency. Furthermore, ach discovered artifact carries within it a wealth of stories, offering limpses into the lives, beliefs, and social structures of past populaions. These tangible remnants of ancient innovation serve as silent itnesses to the continuous human pursuit of precise measurement.

.2 Early Rulers

he earliest rulers, surprisingly, weren't always made of wood or netal. Some of the most ancient examples are simple markings etched nto bone, perhaps a rib from a large animal, or even a piece of careully smoothed stone. These rudimentary measuring devices, often ound in archaeological sites across various continents, reveal a funlamental human need for standardization and precision, even in the nost primitive societies. The markings, usually evenly spaced, repreent the earliest attempts to create consistent units of length, crucial or tasks ranging from building simple shelters to crafting effective unting tools. The irregularity of spacing in some instances speaks to he challenges of consistent manufacturing with limited technology, ut the very existence of these artifacts proves that measurement was concern even before sophisticated technologies appeared.

onsider the Egyptian cubit, a unit of length based on the forearm meaurement. While not a ruler in the strictest sense, its widespread use cross ancient Egypt demonstrates a need for shared measurements for construction and trade. Ancient Egyptian monuments, like th pyramids, represent a remarkable feat of engineering that required extremely precise measurements. While their methods might seem crud to modern eyes, the astonishing accuracy of these structures speaks volumes about their understanding of measurement and the effectivenes of their tools, despite lacking sophisticated instruments. The subtl variations in cubit length across different regions and time periods ar themselves a topic of scholarly study, showcasing the complexities o ancient measurement systems and the difficulty of maintaining standardization over large areas and lengthy time spans.

The development of more advanced rulers occurred gradually, witl the introduction of materials like wood and later, metal. Woodei rulers, often made from hardwoods like ebony or boxwood, allowe for greater precision and durability. The use of standardized units became more common, leading to a greater consistency in measurement across various projects. The creation of these rulers shows a shift in th nature of measuring from a largely intuitive and localized practice towards a more formalized, standardized system. In fact, the very creation of the objects shows the increased level of societal organizatioi necessary to support broader building projects, suggesting a deep connection between technological advancement and the development o more complex societal structures.

Metal rulers, primarily bronze and later iron, marked a further signifi cant advancement. The introduction of metal allowed for greater accuracy in the creation of markings, increased durability, and greater resistance to wear and tear. We see evidence of sophisticated metalworking technologies in these objects, showing a marked improvement in manufacturing capabilities. Moreover, the precision evident in some of these rulers suggests that specialized artisans dedicated to the creation of these instruments existed within ancient societies. These specialized artisans, in turn, suggest a highly developed division of labor and a more intricate socio-economic structure.

These developments were not isolated incidents. The refinement of ruler technology was integral to the advancement of various fields, including architecture, carpentry, and surveying. The standardization of measurement directly impacted trade by providing a common language for exchanging goods and services. The precision afforded by metal rulers would have facilitated the construction of more complex structures and the manufacture of more intricate artifacts. One can directly trace a link between the evolution of these simple measuring tools and the development of more advanced technologies throughout various ancient civilizations. The development of more advanced societies hinges on seemingly minor advancements. The ability to measure accurately and consistently was a key element in this evolution, as it acted as a foundation upon which all other technological advancements could be built. The early ruler was not simply a tool, but a critical catalyst of social and technological progress.

2 Chapter 2: Agricultural Innovation

The humble plow, a seemingly simple tool, represents a monumental leap in agricultural technology. Its evolution, from the rudimentary scratch plow of ancient civilizations to the complex, iron-clad plows of the 19th century, reflects humanity's persistent drive to increase efficiency and yield. Early plows, often crafted from wood and sharpened with stone or bone, required immense human effort. The introduction of metal, first bronze and then iron, significantly enhanced durability and effectiveness, allowing for deeper plowing and the cultivation of heavier soils. Consider the intricate design of the moldboard plow, its curved shape cleverly designed to turn over the soil and bury weeds, a testament to the observational skills and engineering prowess of early farmers. Different plow designs catered to varied terrains and soil types, illustrating a remarkable adaptation to local conditions. The development of the wheeled plow further reduced the physical strain, increasing productivity and efficiency. This progression highlights not just technological advancement, but a profound understanding of soil mechanics and agricultural principles.

The seed drill, a seemingly minor invention, dramatically reshaped agricultural practices. Before its widespread adoption, seeds were typically broadcast by hand, a wasteful and imprecise method. The invention of the seed drill, with its precise seed placement and controlled depth, revolutionized planting, leading to significant increases in crop yields. Early seed drills, often made of wood and employing simple mechanical mechanisms, paved the way for more sophisticated designs incorporating precision engineering principles. These drills enabled efficient planting, reduced seed wastage, and enhanced germination rates. Variations in seed drill design accommodated different seed sizes and planting densities, underscoring the adaptability of this essential tool. Its impact extended beyond mere efficiency; the accurate planting facilitated by seed drills encouraged better crop spacing, optimizing sunlight and nutrient uptake. The evolution of the seed drill, from a simple wooden device to a more complex iron implement, reflects the continuous refinement of agricultural technology.

Reaping tools have also undergone a significant evolution, marking a parallel trajectory with agricultural productivity. The sickle, a simple curved blade, was the primary harvesting tool for millennia. Requiring significant manual labor and skill, it limited harvesting speed and

fficiency. The development of the scythe, with its longer blade and im-
roved ergonomics, significantly increased productivity. The scythe's weeping motion allowed for a more efficient cutting action, reducing he time and effort required to harvest a field. The subsequent arrival f mechanical reapers, powered first by horses and then by steam, rep-esented a true revolution. These machines, with their rotating blades nd automated separation of grain from stalks, drastically reduced he labor required for harvesting. The transition from manual reaping ools to mechanical harvesters dramatically altered agricultural land-capes, allowing for the cultivation of larger areas and the efficient pro-essing of greater volumes of crops. The implications for food produc-ion and societal structures were immense. The evolution of these reap-ng tools demonstrates not just technological progress, but also a con-inuing human endeavor to enhance food security.

2.1 Plows & Harrows

he plow, in its simplest form, is a deceptively simple tool. A sharpened iece of wood or metal, dragged through the earth to turn the soil – et its impact on civilization is monumental. From the earliest wooden rd, a rudimentary scratch plow requiring considerable animal power, o the more sophisticated wheeled plows of the Roman era, the evo-ution showcases incremental improvements driven by necessity and ingenuity. The heavy, iron-clad plows of the Middle Ages, with thei moldboards designed to effectively turn the furrow, represent a signif icant leap forward, enabling deeper tillage and increased agricultura yields. These changes weren't merely technological; they represente shifts in agricultural practices, leading to increased food productio and supporting larger, denser populations. The development of differ ent plow types – the swing plow, the double moldboard plow – reflect regional variations and adaptations to diverse soil conditions. Eac improvement, however small it might seem in isolation, contribute to a greater revolution in farming techniques. Studying the evolutio of the plow is to study the history of human ingenuity's response to th fundamental need for food production.

The harrow, often overlooked in favor of its more prominent agricul tural companion, is equally crucial. Following the plow, the harrow' function is to break up clods of soil, level the field, and prepare th seedbed. Early harrows were likely simple, consisting of branches o wooden beams with embedded stones or sharpened sticks. These rudi mentary designs were effective, yet limited in their ability to thor oughly refine the soil. Over time, however, the construction evolved Iron tines replaced the earlier, less durable materials, leading to mor effective and efficient soil preparation. The introduction of differen harrow designs – bush harrows, smoothing harrows, spring-tooth har rows – each catered to specific needs and soil types. The developmen of these different designs demonstrates a subtle yet vital understanding of soil mechanics and the needs of different crops. The evolution of the harrow provides a fascinating insight into the development of precision in agricultural practices, a move toward greater control over the land and increased efficiency.

Consider the impact of these implements on societal structure. The adoption of more efficient plows and harrows directly correlated with increased agricultural production. This surplus, in turn, fueled the growth of settlements, enabled specialization of labor, and facilitated the rise of more complex societal structures. The relative ease of working the land, facilitated by these improved tools, freed up a portion of the population from purely agricultural pursuits, leading to advancements in other fields like manufacturing, trade, and the arts. This interplay between technological advancement and societal evolution highlights the profound consequences of seemingly minor technological innovations. The interconnectedness of these developments is undeniable, demonstrating the impact of innovation on the broad sweep of human history.

The differences between various regional plow designs are quite telling. Consider, for instance, the variations in plow design between the wet, heavy soils of Northern Europe and the drier, lighter soils of the Mediterranean. These variations reflect not only differing soil conditions but also the ingenuity of local blacksmiths and farmers who adapted their tools to suit their environment. The materials used – locally sourced wood, iron, or even bronze – further highlight the importance of resourcefulness and adaptation in tool development. Studying these regional variations provides valuable insights into local economies, material availability, and the technological exchange networks that connected different communities. Understanding this geographical and cultural nuance is crucial for a comprehensive appreciation of the plow and harrow's story.

Acquiring antique plows and harrows today requires a keen eye and a knowledge of regional variations. The materials used, the construction techniques, and even the style of the handles can reveal important details about the tool's origin and age. Authenticity is crucial in collecting, and a careful assessment is vital to avoid forgeries or misrepresented items. Researching the history of specific designs, consulting with other collectors, and examining detailed photographs can help ensure a discerning approach. The beauty lies not just in the tools themselves, but in the story they tell of human resourcefulness, technological innovation, and the ongoing quest to cultivate the land. The reward is not just possession, but a tangible connection to the past and a deeper understanding of agricultural history.

2.2 Seed Sowers

Early seed drills, often overlooked in favor of the more dramatic plow, represent a fascinating intersection of agricultural necessity and inventive engineering. Consider the simple act of sowing seeds; seemingly straightforward, yet historically fraught with inefficiency. Scattering seeds by hand led to uneven distribution, wasted seeds, and ultimately, reduced yields. The development of seed drills, therefore, wasn't just about planting—it was a significant step towards optimizing agricultural production, a leap forward that profoundly influenced food security and economic prosperity.

These early implements varied widely in their design, reflecting the diverse agricultural landscapes and technological capabilities of different regions. Some were remarkably rudimentary, little more than modified boxes with holes, while others showcased impressive craftsmanship and mechanical ingenuity. Wooden seed drills, for instance, relied on simple rotating mechanisms to distribute seeds, showcasing resourcefulness using readily available materials. More advanced versions incorporated gears and even wheels, ensuring a more consistent and controlled planting process. The evolution of these devices can be seen as a microcosm of technological progress, with each refinement aiming for greater efficiency and precision.

The impact of the seed drill extended beyond simply increasing crop yields. The precise planting facilitated by these tools enabled farmers to optimize planting distances, leading to better growth and resource utilization. This resulted not only in increased yields but also in improved crop quality. Moreover, the increased efficiency of seed drills meant that farmers could cultivate larger areas of land, contributing to societal changes such as increased food surplus, population growth, and the rise of more complex economies.

Imagine the careful planning required to craft a seed drill. Each component, from the seed hopper to the distributing mechanism, needed to be precisely designed and crafted to function correctly. The selection of materials was crucial, requiring an understanding of durability, resistance to weathering, and ease of fabrication. The construction itself was a testament to the skill and craftsmanship of the individuals who created them, often working with limited tools and resources. Their dedication is evident in the surviving examples, many of which are beautifully crafted and surprisingly robust.

But the seed drill was more than just a tool; it was a symbol of human ingenuity and adaptation. It represented a sustained effort to improve upon existing methods, a relentless pursuit of better agricultural practices. It showcases humanity's ability to solve fundamental problems through innovation and careful engineering. The story of the seed drill is thus not simply a technical one, but a narrative of human resilience, adaptation, and the persistent desire for improvement. It is a comelling story of how human ingenuity transforms agriculture, driving rogress through practical innovations.

Many early seed drills were specifically adapted to the needs of the re-ion and the crops being planted. Different seed sizes and soil condi-ions required unique design elements. For instance, a seed drill de-igned for planting wheat in heavy clay soil would differ significantly rom one intended for sowing barley in lighter, sandier conditions. he ingenuity of these designs is reflected in their adaptability, high-ighting the relationship between technological innovation and agri-ultural practices. This deep connection between tools and the specific nvironment illustrates the profound and complex interplay between echnology and the environment.

he construction of these seed drills varied according to available re-ources and technological advancements. While early examples were argely hand-crafted from wood and basic metal components, later lesigns incorporated more sophisticated materials and manufactur-ng techniques. This transition mirrors the broader advancements in netalworking and engineering that occurred during this period. The hift from predominantly wooden tools to metal components reflects he technological developments that occurred over time, which led to tronger, more durable, and more precise tools.

Beyond their immediate agricultural benefits, seed drills provide valu-able insights into the social and economic contexts in which they were developed and used. They offer a glimpse into the daily lives of farm ers, the challenges they faced, and their innovative responses. The de velopment of seed drills can be seen as a testament to the ingenuity an resourcefulness of people in response to the challenges of food produc tion. The survival of many of these implements through the centurie speaks to their importance and provides valuable insights into the his tory of technology and culture.

Furthermore, the seed drill embodies a timeless principle: the huma drive for efficiency. It reveals humanity's ongoing quest to optimiz processes, to maximize output with minimal effort. This principle tran scends specific historical contexts and technological advancements. I is reflected in the inventions and innovations across numerous fields demonstrating a fundamental aspect of human nature and the drivin force behind technological progress. The seed drill serves as a powerfu example of how simple ingenuity can make huge impacts on society and how innovation is continuous.

Studying these ancient seed drills offers far more than just a histori cal perspective; it offers valuable lessons for contemporary agriculture In today's world, which faces the challenges of growing population and climate change, understanding the innovations of the past, like th seed drill, is crucial to advancing sustainable agricultural practices fo the future. By understanding these earlier techniques we can better ap proach innovative solutions to address modern challenges, reflectin on the past to benefit the future. The ingenuity displayed in their creation remains an invaluable lesson for today's innovators.

2.3 Reaping Tools

The sickle, a seemingly simple tool, holds a significant place in agricultural history. Its curved blade, crafted from flint, bronze, or later, steel, allowed for the efficient harvesting of grains. Early sickles, often depicted in ancient artwork, were relatively small and handheld, requiring considerable manual dexterity and physical strength. Their effectiveness depended heavily on the skill of the user, and the yield per hour would be comparatively low, a testament to the immense effort involved in feeding populations before mechanization. The evolution from these early tools to more advanced reaping implements reveals a fascinating narrative of agricultural ingenuity and technological progress. It's a journey from hand-held tools to collaborative efforts to maximize harvest potential. The introduction of the cradle scythe, with its longer blade and attached gathering mechanism, marked a significant leap forward. This innovation allowed for a greater swath to be cut in each movement, considerably increasing efficiency. The cradle's design, while relatively simple, represents a crucial step in harnessing human power more effectively for agricultural production. This progression from individual effort toward more efficient team-based work patterns is a story that reflects broader societal shifts.

Consider the advancements in materials and the impact on these tools. The transition from flint to bronze, and finally to steel, significantly enhanced the durability and sharpness of the blades. Steel, with its superior strength and resilience, allowed for longer, sharper blades, furthering the efficiency of harvesting. The improved durability also meant less frequent tool replacement, reducing downtime and contributing to overall productivity. This isn't simply about the materials; it speaks to the development of metallurgy and its transformative effect on various aspects of daily life, not just in agriculture. This interconnectedness between technological advancements in unrelated fields underscores the importance of looking at tools not in isolation, but as products of broader cultural and technological developments. The use of materials reflects the knowledge and skills of the time, and therefore, provides a rich lens through which to explore the social and economic context of their creation and use.

The development of reaping hooks, with their shorter, more robust blades, further refined the harvesting process. These were often used for cutting closer to the ground, ensuring that no grain was left behind. This level of detail demonstrates the ongoing drive for optimization and the understanding of the importance of minimizing waste and maximizing yield. The design itself, reflecting the need to balance the cutting effectiveness with the ergonomic needs of the user, makes it more than just a tool; it's a testament to the observation of practical needs in tandem with mechanical efficiency. We might take these tools for granted, but observing these implements under a careful lens demonstrates a deep understanding of the user needs, and of human factors that are relevant also today, in the designing of modern tools.

Moving beyond hand-held implements, we encounter the evolution of more complex mechanical reaping machines. These involved ingenious mechanisms for cutting, gathering, and even binding the harvested grain. Early examples incorporated gears and levers, demonstrating a growing sophistication in mechanical engineering, while the use of horses or oxen demonstrated another level of human creativity and problem-solving. These advancements dramatically transformed agricultural practices, allowing farmers to harvest far larger areas in a fraction of the time. These machines are not simply improvements, but fundamental shifts in the relationship between human labor and technological solutions. The shift represents a paradigm change in agricultural practices, which had profound consequences for society, stimulating population growth and trade networks on a scale previously unimaginable.

Ultimately, the evolution of reaping tools represents a fascinating microcosm of technological progress. It showcases the continuous drive to improve efficiency, optimize techniques, and better harness available resources. From the humble sickle to the complex mechanical reapers, each innovation reflects an understanding of human needs and the possibilities of technical ingenuity. These tools are not just artifacts of the past, they are tangible testaments to the enduring human spirit of innovation, providing valuable insight into the social, economic, and technological forces that shaped human history. The story of these tools teaches us that the most significant inventions are often those that address fundamental human needs and that improvement is a gradual, incremental process involving countless innovations, each adding a small amount to an overall better and more efficient system of work.

3 Chapter 3: Precision Engineering

he development of precision engineering is intrinsically linked to the dvancement of human ingenuity. Consider the humble clock. Early imekeeping devices, while functional, lacked the accuracy achievable vith later precision-engineered mechanisms. The transition from rudi-nentary sundials and water clocks to intricate clockwork mechanisms epresents a pivotal moment, demanding a level of craftsmanship and nderstanding of mechanics previously unseen. This shift necessitated he creation of specialized tools capable of producing parts with toler-nces far beyond anything imaginable in previous eras. The precision equired for gears, springs, and escapements fueled innovation in ma-hining techniques and the development of increasingly sophisticated neasuring instruments.

Aicrometers and calipers, seemingly simple tools, were revolutionary. heir ability to measure to hundredths or even thousandths of a millimeter allowed for a level of accuracy that transformed manufactur ing. Before their advent, reliance on hand-eye coordination and les precise instruments frequently led to parts that simply wouldn't fit cor rectly, resulting in rejected work and costly mistakes. The introductio of these tools allowed for the creation of interchangeable parts, a con cept that would become fundamental to mass production and the as sembly line. Their evolution, from simple designs crafted by hand t finely engineered instruments, speaks to the relentless pursuit of im proved accuracy. One can imagine the countless hours of painstakin effort required to manufacture each component, illustrating the leve of skill and patience demanded of these early precision engineers.

The lathe, a cornerstone of mechanical engineering, stands as testa ment to the human desire for creating perfectly shaped components From simple wood-turning lathes to the complex metal-working ma chines of the Industrial Revolution, this machine has undergone con siderable evolution. The development of precision lathes, capable o holding incredibly tight tolerances and producing intricately shape parts, was a significant technological leap. These lathes, equipped wit specialized chucks, rests, and cutting tools, enabled the creation of com ponents with exquisite detail. The development of the lathe, therefore isn't simply the story of a machine's evolution; it's a testament to th ongoing quest for ever-greater manufacturing accuracy. The skille machinists who operated these lathes possessed remarkable dexterity knowledge, and patience. Their craft involved intricate adjustments and a deep understanding of the materials they worked with.

Gear-cutting tools represent another significant achievement in precision engineering. Before the development of dedicated gear-cutting machinery, the creation of gears was a painstaking process, often involving laborious hand filing and shaping. The introduction of specialized gear-cutting machines revolutionized gear production, enabling the creation of accurate gears with consistent tooth profiles. This resulted in smoother running mechanisms, reduced wear, and increased reliability – critical in the development of complex machines like clocks, looms, and eventually, automobiles. The design and construction of these machines themselves represented a peak in the precision engineering of their time, showcasing the interconnectedness of technological advancement. The very process of creating these tools reflected the level of precision already achieved in the field.

Finally, advanced gauges represent the culmination of centuries of refinement in measurement techniques. These tools, ranging from simple comparative gauges to highly sophisticated instruments measuring minute variations, are essential for quality control in manufacturing. The need for consistency and reliability in production led to the development of sophisticated gauges that could ensure manufactured parts conformed to exacting standards. This focus on quality control was crucial in enabling the development of more complex and dependable machinery. The evolution of these gauges, from simple go/no-go gauges to electronic measuring instruments, reflects the constant push for even higher levels of precision in manufacturing. The development of such precise measurement devices further empowered the development of the intricate machinery capable of manufacturing them. This iterative process is a testament to the power of human ingenuity and the relentless pursuit of perfection.

3.1 Clockwork Mechanisms

The rhythmic tick-tock of a clock, a seemingly simple sound, belies the intricate mechanisms within. Clockwork, far from being merely a method of timekeeping, represents a pinnacle of precision engineering, a testament to human ingenuity and a crucial stepping stone in the development of modern technology. Understanding its history unveils not only the evolution of timekeeping but also the development of manufacturing techniques that underpinned the Industrial Revolution.

Early clocks, often massive and unwieldy, relied on rudimentary escapements and weights, their accuracy limited by the friction inherent in their design. These early examples, housed in grand cathedrals or stately homes, were more than mere timekeepers; they were statements of power and wealth, testaments to the skill of their creators. The craftsmanship involved, a blend of artistry and engineering, is astonishing, each component meticulously crafted and assembled.

The transition to smaller, more portable clocks involved significant advancements in gear design and material science. The development of finer materials, such as brass and steel, allowed for the creation of smaller, more precise gears, thus improving accuracy and reducing wear. Simultaneously, the invention of the pendulum, with its inherent isochronism, revolutionized timekeeping, leading to significantly more accurate clocks. The pendulum, a seemingly simple device, provided a consistent and reliable regulator for the clock's mechanism.

However, even with the pendulum, maintaining accuracy remained a challenge. Temperature fluctuations, for example, could subtly alter the length of the pendulum, affecting its period and consequently the clock's timekeeping. The development of temperature compensation mechanisms, ingenious devices that adjusted the pendulum's length in response to temperature changes, represented a further leap forward in precision. These mechanisms, often intricate and elegantly designed, highlight the importance of adapting technology to its environment.

The creation of the balance spring and hairspring, which regulate the timing of portable clocks and watches, marked another crucial development. These tiny components, requiring exceptional skill to manufacture, enabled the creation of timepieces small enough to be worn on the wrist. The accuracy achievable with these springs surpassed that of earlier pendulum clocks, opening the door to a new level of precision timekeeping.

The evolution of clockwork mechanisms also had a profound impact on the development of manufacturing techniques. The demand for precision components, such as gears and springs, spurred innovation in metalworking techniques, leading to advancements in machining, casting, and other essential processes. The need for interchangeability, allowing for easy repair and replacement of parts, further encouraged standardization and the development of mass-production techniques.

The intricate mechanisms found in antique clocks and watches represent not only a fascinating historical record of timekeeping but also a window into the remarkable craftsmanship and technological advancements of past eras. Each gear, spring, and escapement, precisely engineered and meticulously assembled, speaks to the dedication and skill of the artisans who created them. Studying these mechanisms offers valuable insights into the history of technology and the evolution of manufacturing techniques, reminding us of the profound impact of even seemingly simple inventions.

Collecting antique clockwork mechanisms offers a unique opportunity to appreciate these fascinating artifacts. Beyond their aesthetic appeal, these objects offer a tangible link to the past, a connection to the individuals who designed, manufactured, and used them. Each piece tells a story, a silent narrative of human ingenuity and the relentless ursuit of precision. Their study not only enhances our understanding f the history of timekeeping but also sheds light on the wider context f technological advancement and the interconnectedness of diverse ields of human endeavor. The intricate dance of gears, the delicate in- erplay of springs, and the rhythmic pulse of the escapement — these lements combine to create not just a timekeeping device, but a testa- nent to the enduring human fascination with precision and the relent- ess pursuit of perfection. From the monumental astronomical clocks f past centuries to the exquisitely crafted pocket watches of a later ra, each artifact represents a chapter in the ongoing story of human nnovation. Examining these intricate devices is not only about under- tanding their mechanics but also about appreciating the artistry and ngenuity behind their creation. These antique pieces are not just me- hanical devices; they are historical records of technology and testa- nent to human skill.

3.2 Micrometers & Calipers

Micrometers and calipers, seemingly simple instruments, represent a iant leap in the precision required by burgeoning industries of the 9th and 20th centuries. Their development wasn't a singular event, but ather a gradual refinement of earlier measuring techniques, build- ng upon centuries of accumulated knowledge and the ever-increasing demands for accuracy in manufacturing. Think of the intricate clock work mechanisms discussed in the previous chapter; their creation de manded tools capable of measuring minute differences, a challeng met by the gradual evolution of these precise instruments.

The earliest forms of these devices were far cruder than their mod ern counterparts. Simple calipers, often fashioned from wood or bone acted as rudimentary dividers, allowing craftspeople to measure th diameter of objects or the distance between two points. Accuracy wa limited, dependent upon the skill and eye of the user and the inheren limitations of the materials used. Yet, they served their purpose, pro viding a level of precision previously unattainable through purely vi sual estimations. These early forms, unearthed in archaeological dig or preserved in museum collections, provide a fascinating glimpse int the evolution of measurement. Analyzing these tools, we can see th ingenuity of their creators, working with limited materials to achiev remarkable results.

The invention of the micrometer screw gauge revolutionized precisio measurement. This ingenious device, employing a precisely threade screw to achieve fractional measurements, dramatically increased th accuracy attainable. Instead of relying on human eyesight to judg minute differences, the micrometer provided a mechanical means o measurement, reducing the margin for error. Early micrometers wer often made of brass or steel, their construction reflecting the high leve of craftsmanship demanded for their production. Examining antique micrometers, one can appreciate the meticulous attention to detail exhibited by their makers, the delicate balance between strength and precision clearly evident in their design. These weren't mass-produced items; each was a testament to skilled engineering.

However, the micrometer's accuracy wasn't always perfect. Factors like temperature fluctuations and wear on the screw could affect its reliability. These inherent imperfections highlight the ongoing quest for better measuring techniques, a constant drive to improve precision in manufacturing. This continuous development isn't just a story of technological advancement but also a narrative about the human desire to refine, improve and achieve greater accuracy in our work. Early users understood and accounted for these limitations, employing techniques such as multiple measurements and careful calibration to achieve the required level of accuracy.

The development of calipers mirrored that of micrometers, progressing from simple wooden or metal dividers to more sophisticated designs incorporating vernier scales or micrometer screws. These enhancements increased the precision achievable, allowing craftspeople to make finer adjustments and create more intricate components. The introduction of steel construction improved durability and resistance to wear, further enhancing their longevity and accuracy. The evolution of these instruments is directly connected to the advancements in manufacturing processes, particularly in the rise of mass production. These tools didn't just measure; they became essential instruments in making precision manufacturing a reality. In fact, the demand for precision in the manufacturing industries fueled further refinement in the design and manufacturing of calipers and micrometers themselves.

Antique micrometers and calipers are not just tools; they are artifacts of technological history. They reflect the ingenuity of past engineers and craftsmen, the constant drive for improvement, and the ever-increasing demand for accuracy in manufacturing. Collecting and studying these instruments provides a valuable insight into the development of precision engineering and the history of industrial technology. Their detailed construction, often bearing the maker's marks or the stamps of their manufacturers, is a fascinating record of craftsmanship and innovation. Their preservation is crucial, not only for understanding the past, but for appreciating the foundations upon which modern precision engineering is built. Each tool tells a story; a story of the human desire to achieve the impossible, to refine and perfect our understanding of the world around us. To the discerning collector, an antique micrometer or caliper isn't just a piece of metal; it's a window into the past, a testament to human innovation, and a remarkable relic of a technological evolution that continues to this day.

3.3 Lathes & Drills

The rhythmic whir of a lathe, the precise bite of a drill—these sounds, once commonplace in workshops worldwide, now evoke a sense of history. The lathe, a machine for shaping wood or metal, boasts an ancestry stretching back millennia. Early versions, simple devices employing friction and hand-cranks, gradually evolved into the complex, precision instruments we associate with the Industrial Revolution. Consider the ingenious adaptations: the use of different woods for different lathes, the evolution of the bed to accommodate larger pieces, and the incorporation of gears and levers for finer control. This evolution wasn't just a matter of increased power; it reflects a growing understanding of materials science and mechanical engineering. Each refinement, each innovation, represents a significant step in the capability of the machine itself and what could be produced with its aid. The development of the screw-cutting lathe, for example, allowed for the mass production of standardized parts, a development that profoundly impacted everything from clockmaking to shipbuilding.

Focusing on the drill, its seemingly simple operation belies a rich history of innovation. Hand drills, some dating back to ancient Egypt, utilized the principles of leverage and friction to bore holes. Their evolution is marked by increasing efficiency and precision. The brace and bit, for example, offered greater control and torque, enabling the creation of larger and deeper holes. These tools were essential for carpenters, cabinetmakers, and other artisans. Even the seemingly minor advancements, such as improved bit designs or the incorporation of interchangeable chucks, dramatically altered their capabilities. The introduction of power drills, powered by steam or electricity during the 19th century, further revolutionized drilling, allowing for greater speed, depth, and accuracy, enabling engineers to complete projects they previously could not.

The relationship between lathes and drills is far more significant than their individual evolution suggests. They were often used in conjunction, complementing each other in the creation of complex objects. A lathe might be used to shape a piece of wood or metal, and then a drill would be employed to create holes for screws, pins, or other components. This synergy accelerated the development of both tools. The increased precision of lathes led to demands for more accurate drills, and vice versa. This symbiotic relationship highlights the interconnected nature of technological advancements. One invention often drives progress in others. The creation of better drills enabled more intricate lathe work, furthering the development of both, a clear sign of the importance of interconnected industrial advancement and a driving force of technical progress. This process is an excellent example of how technological developments do not happen in isolation, but are often mutually reinforcing.

onsider the materials used in the construction of these tools themselves. Early lathes might have been crafted from simple wood, later nes incorporated iron and steel for increased durability and precision. imilarly, drills evolved from softer woods to tougher metals, again eflecting improvements in both manufacturing processes and mateials science. The metallurgy of different tools and different times relects both manufacturing technology and the materials available at he time. These advancements allowed for ever greater accuracy and eficiency in both tools. The very materials from which these implements vere made reveal a story in themselves; they speak of technological caability, resource availability, and the ingenuity of their creators.

he aesthetic qualities of antique lathes and drills are frequently overooked, but they represent another important aspect of their study. The raftsmanship of many of these tools is remarkable, demonstrating level of skill and artistry that is often absent from mass-produced nodern equivalents. From the carefully turned wood of a hand-crank athe to the elegant curves of a brass-handled drill, these objects are not ust functional tools; they are works of art, their design is functional nd artistic at the same time. Their aesthetic appeal contributes signiificantly to their collector value, further increasing their desirability mongst collectors and history enthusiasts. Studying these tools helps is understand not just the technological advancements of the past but lso the artistic sensibilities of the people who crafted them. This appreciation is essential for a comprehensive understanding of the items context.

Collecting antique lathes and drills is not simply an exercise in acquiring rare objects; it's a journey into the history of craftsmanship manufacturing and engineering. These tools are tangible links to past generations of artisans and engineers, each bearing silent witness to countless hours of skilled labor and innovative design. The preservation of these tools is, therefore, a responsibility. By studying them, we can gain a deeper appreciation for the ingenuity of past generation and the foundational principles upon which modern manufacturing i based. This understanding is invaluable not only for collectors and historians but for anyone interested in the evolution of technology and the ingenuity of human invention. The study of these objects is an enriching experience, connecting us to a history filled with both practicality and artistry.

3.4 Gear Cutting Tools

Gear cutting, a seemingly simple process, reveals a fascinating history of intricate tools and ingenious techniques. Consider the earliest methods: rudimentary files and hand-made tools, painstakingly shaping individual teeth, a process demanding immense skill and patience. Each tooth, a testament to the craftsman's dedication, was individu ally shaped and refined, a laborious task requiring significant time and precision. The resulting gears, though functional, were often limited in accuracy and consistency. Early examples were mostly found in clocks and other time-measuring devices, showcasing the importance of accurate gear cutting even in its nascent stages.

The development of specialized gear cutting tools marked a significant leap forward. Imagine the introduction of the indexing head, a device enabling precise rotation of the workpiece, allowing for the creation of uniformly spaced teeth. This innovation drastically reduced production time and vastly improved the consistency of the finished product. The indexing head, often part of a larger lathe setup, allowed artisans to create gears with far greater accuracy than previously possible. This crucial advancement enabled the mass production of gears for a range of applications beyond horology, ushering in an era of more complex machinery.

Further advancements involved the creation of specialized cutting tools themselves. No longer were simple files sufficient. The advent of purpose-built cutters, crafted from hardened steel, allowed for faster and more precise cutting of gear teeth. These cutters, often shaped to match the desired tooth profile, could rapidly produce accurate and consistent results. The development of specific cutter geometries, like involute profiles for improved meshing, further enhanced efficiency and reduced wear. The selection of the right cutter material, for instance, high-speed steel, was also crucial to prolonging the tool's lifespan and achieving higher quality results. This resulted in a significantly improved quality and wider usage of gears in various industries. The next evolution involved the introduction of shaping and hobbing machines. These automated tools significantly reduced the manual labor involved in gear cutting. Shaping machines utilized a reciprocating cutter to shape the teeth, resulting in a smoother and more refined finish. Hobbing machines, however, revolutionized the process entirely, utilizing a rotating cutter to generate the entire gear profile in a single pass. These machines could efficiently and quickly produce high-quality gears with unprecedented accuracy, further driving down production costs and opening up new possibilities for various engineering applications. The precise control over the hobbing process offered immense flexibility, adapting to various gear sizes and configurations. Even with automated processes, the selection and maintenance of cutting tools remained critical. Proper sharpening and lubrication were crucial for maintaining cutting efficiency and prolonging the lifespan of these often-expensive tools. Regular inspection for wear and tear was also paramount. This preventative maintenance not only guaranteed high-quality output but also prevented costly downtime. The use of specialized lubricants, adapted to the specific metals and cutting speeds, were also essential components, minimizing friction and maximizing performance.

Beyond the mechanics of cutting, the understanding of gear geometry remained crucial. The development of precise measuring instruments, like micrometers and verniers, played a key role in ensuring the accuracy of gear production. The careful calculation and precise execution of critical dimensions, including module, pressure angle, and tooth profile, ensured that gears meshed correctly. This attention to detail resulted in efficient power transmission, reducing wear and tear on the entire system.

The study of gear cutting tools provides an insight into the evolution of manufacturing techniques. From rudimentary hand tools to automated machines, this progression reflects the constant drive towards efficiency and precision. The improvements in gear-cutting techniques significantly impacted various industries, enabling the development of more complex and efficient machinery across diverse applications, from clocks to automobiles, to industrial machinery. The evolution of gear cutting is a testament to human ingenuity and the continuous quest for improving both the precision and efficiency of manufacturing techniques. Each generation of tools built upon the previous ones, incorporating new materials, designs, and manufacturing processes, demonstrating a clear understanding of engineering principles. The legacy of these tools, many of which are considered antique today, lives on in the precise and reliable mechanisms found in countless machines across the world. The collection and study of these tools offer a tangible connection to the past and an appreciation for the foundational innovations that shaped modern engineering.

3.5 Advanced Gauges

Advanced gauges represent a pinnacle of precision engineering, surpassing the capabilities of simpler measuring tools discussed previously. These weren't merely tools; they were statements of technological prowess, reflecting the increasing demand for accuracy in manufacturing and quality control. Their intricate designs and precise functionalities reveal much about the ingenuity and dedication of the artisans and engineers who crafted them. Think of them as the ultimate expression of the craftsmanship explored in earlier chapters.

Consider, for example, the Bourdon tube pressure gauge. Its elegant simplicity belies its ingenious mechanism. A flattened, curved tube, sealed at one end, straightens slightly when pressurized. This minuscule movement is amplified via a series of gears and levers, connected to a dial that displays the pressure. The accuracy of these gauges, especially the finer ones, depended heavily on the materials used – the elasticity of the tube, the precision of the gears – and the skill of the maker in assembling the device. Variations in design catered to different pressure ranges, from low-pressure applications in delicate instruments to the immensely high pressures found in early steam engines. Collecting hese gauges provides a fascinating glimpse into the development of ressure measurement technology across various industrial sectors.

eyond pressure, there was a constant need for precise measurement f other physical properties. The development of dial indicators, for nstance, demonstrated a move toward even greater accuracy. These evices, often used in conjunction with micrometers and calipers, al- owed for incredibly fine measurements of linear displacement. Their pplication spanned various fields – from assessing the wear of ma- hine parts to precise adjustments in manufacturing. A key element in heir design was the lever system, transferring the minute movement f a contact point into a significant deflection on the dial. Collectors vill find subtle yet significant differences in the design of these indi- ators; examining these variations provides a window into manufac- uring techniques of the time. The materials – often hardened steel and inely crafted gears – speak volumes about the engineering standards xpected, and achieved.

urthermore, the evolution of thread gauges highlights the increasing mportance of standardization in manufacturing. Before widespread doption of standardized thread sizes, each manufacturer might have sed slightly different dimensions. This created compatibility issues nd significant frustration. The introduction of thread gauges – sets f precisely made components that allowed for checking the pitch and iameter of screws and bolts – represented a major step toward interchangeability and streamlined production. These tools were, and re main, an essential part of any machinist's arsenal, a testament to th desire for efficiency and uniformity in industrial processes. Examinin the variety of thread types represented in a collection of gauges show the diverse technologies and applications they served.

Another fascinating category within advanced gauges are the special ized instruments developed for specific industries. For instance, the tex tile industry relied heavily on gauges for measuring thread count an fabric thickness. Similarly, the paper industry employed gauges to de termine the thickness and weight of paper sheets. These specialize tools show us the degree to which measurement was integrated int various industrial processes and how the unique needs of individua sectors drove the development of specialized metrology equipment. Th differences in designs across sectors highlight the interaction betwee technological progress and the demands of specific industries.

Finally, the collection and study of these advanced gauges offers deeper understanding of the relationship between design, manufac ture, and industrial progress. The quality of the gauge often reflecte the quality of the products it was used to measure, reflecting the im portance of precision in all levels of production. Analyzing materials construction, and the evolution of designs throughout the history o these implements will give the collector a deeper appreciation for th importance of accurate measurement in shaping our industrial world

The elegance of their designs, coupled with their practical significance, makes advanced gauges a compelling area of study for collectors interested in precision engineering.

4 Chapter 4: Household Artifacts

Early irons, far from being mere heating devices, represent a fascinating evolution of technology and societal needs. Consider the simple hand-held smoothing iron, its weight and design reflecting the laborious process of flattening wrinkles. These early examples, often crafted from cast iron, demonstrate a remarkable understanding of heat distribution and material properties. The development of these tools mirrored the changing fabrics and clothing styles, with adjustments in weight, shape, and even the addition of handles for improved ergonomics. The transition to heavier, stove-heated irons marks a significant leap, demonstrating the increasing integration of household tools with evolving fuel sources and domestic architecture. The intricate details, such as the careful placement of vents or the shape of the iron's base, all contribute to a deeper understanding of both technological advancement and the daily lives of those who used them. These subtle differences tell stories of resourcefulness and adaptation, reflecting the ingenuity of the makers and the needs of the users. From the simplest, hand-held design to the more complex stovetop models, each iron reflects the historical period in which it was made. Studying their construction and stylistic details provides valuable insights into craftsmanship, material availability, and changing cultural preferences. Their evolution mirrors a broader technological advancement, showing how a simple household tool reflects larger societal and economic changes over time. Each example holds a unique story, providing a window into the past.

The butter churn, a seemingly simple device, reveals a surprising amount about the ingenuity and resourcefulness of our ancestors. Early designs, often rudimentary wooden vessels, relied on manual agitation to separate butterfat from milk. The basic design – a container with a paddle – evolved over time to increase efficiency. Some featured ingenious mechanical elements, such as rotating paddles or intricate internal structures designed to maximize churning efficiency. The materials employed – wood, stone, or even metal – also reflected the resources available in different regions and periods. Beyond simple functionality, however, butter churns also represent a fascinating case study in material selection and engineering, often reflecting regional variations and adapting to local conditions. The evolution of these devices highlights the importance of food preservation in pre-

efrigeration societies, and how human ingenuity found ways to im-
rove everyday tasks through experimentation and refinement. The onstruction often involved complex carpentry techniques, reflecting a evel of skill and precision often overlooked. Examining the construc-ion methods allows one to understand not only the technology of the ime but also the craftsmanship that went into its creation. Each but-er churn, with its specific design and construction details, therefore be-omes a microcosm of its time and place.

he evolution from simple hand-powered versions to later designs in-orporating gears and other mechanical features reflects not only tech-ological advancements but also a changing understanding of effi-iency and work optimization. The diversity in design also points to re-ional variations and preferences, showcasing the adaptability of the echnology to local contexts. Furthermore, the decoration and orna-nentation of some churns show a deep connection between functional-ty and aesthetics in daily life. Many churns are not only functional ar-ifacts, but also visually engaging objects demonstrating the artisan's kill and artistic sensibilities. Understanding these elements is key to ınderstanding the wider cultural significance of the objects, showing he intersection of technology, art, and daily life. These butter churns ıre more than just tools; they are testaments to human resourcefulness ınd ingenuity. Their survival today offers a tangible connection to past enerations and a window into their lives. Studying them provides an enriching insight into technology and its social context.

4.1 Early Irons

The earliest irons, far from the sleek electric models of today, offer a fascinating glimpse into the daily lives and resourcefulness of past generations. Imagine the weight of a heavy, cast-iron smoothing iron, its surface often rough and uneven, held in hands calloused from years of labor. These weren't merely tools; they were extensions of the individual, their unique marks and wear patterns a testament to countless hours spent pressing linen, smoothing silks, or meticulously shaping the family's clothing.

Their construction reveals much about the technological capabilities of the time. Many early irons were made of cast iron, requiring skilled foundry work to achieve a consistent shape and weight. The heating process varied, some relying on embers placed inside a hollowed-out body while others were heated directly on a stove or hearth. The simplicity of the design, however, belies the ingenuity involved in their creation. Careful consideration of weight distribution was crucial for optimal usability; an iron too heavy would be unwieldy, while one too light would be ineffective. The handle, often made of wood, was a significant design element, requiring careful shaping to provide a comfortable and secure grip during extended use.

The evolution of the design itself is a compelling narrative. Early iterations were often simple, boxy shapes, reflecting the limitations of casting techniques. As technology improved, so too did the irons' form. More refined shapes emerged, offering better maneuverability and heat distribution. The addition of a handle to the top, rather than the side, signified a step towards ergonomic design. Later models would incorporate features like replaceable handles to prolong their lifespan, reflecting an increasing awareness of both utility and cost-effectiveness. The surfaces, initially smooth or with basic texturing, eventually evolved to include more complex designs, possibly incorporating decorative elements that reflected the social status of the owner.

While seemingly mundane, the presence of a smoothing iron in a household spoke volumes. It signified a level of prosperity that allowed for the maintenance of fine clothing, reflecting social standing and cultural values. The careful pressing of garments was a visible sign of respectability, cleanliness, and social refinement. Moreover, owning a well-maintained iron implied a degree of personal investment and care in one's possessions – a testament to thrift, practicality, and a pride in one's work.

The study of early irons extends beyond simple functionality. They become tangible links to the past, offering historians a glimpse into material culture, technological advancement, and daily life. Analyzing the materials used, the manufacturing techniques employed, and the overall design reveals a wealth of information about the economic, social, and technical environment in which they were produced. The discovery of specific markings on the iron, such as maker's stamps or symbols, can further illuminate their origin and history, allowing for more precise dating and geographic location. It's a journey that unites material artifacts with historical analysis, illuminating the stories of innovation and everyday practicality from the people who used them.

These seemingly simple tools, found in attics, antique shops, and museums, are more than just metal and wood. They are artifacts that tell a story of ingenuity, adaptation, and the enduring human desire for order and neatness. Their preservation and study allows us to appreciate not just the object itself, but the skill, resourcefulness, and craftsmanship behind it. Every scratch, every dent, every imperfection becomes a mark of history, a whisper from the past that connects us to those who came before. The weight, the texture, the design – all offer valuable clues to understanding the lives of generations past. These early irons provide a unique and intimate connection to a bygone era, enabling us to touch the past and understand the significance of everyday objects in shaping human history. Studying them allows us to understand not just the tool, but the culture it served, the people who used it, and the broader context of technological and societal evolution. The stories embedded within these simple implements are far richer than their appearance might suggest; they offer a powerful reminder of the enduring human ingenuity.

4.2 Butter Churns

The rhythmic churn of a butter churn, a seemingly simple device, speaks volumes about the ingenuity and resourcefulness of past generations. Its evolution, from the earliest rudimentary vessels to the more complex designs of the late 19th century, reflects a fascinating interplay of materials science, engineering principles, and the ever-present need for efficient food preservation. Early examples, often carved from wood or fashioned from readily available materials like clay, relied on the basic principle of shaking or rocking to agitate cream and separate the butterfat. These early versions, often passed down through families, bear testament to a time when practicality and longevity were prized above all else. The design, while simple, reveals a deep understanding of the physics involved; the rhythmic movement creating the friction necessary for the transformation of cream into butter.

Consider the transition from simple hand-powered churns to more sophisticated mechanisms. The introduction of a dasher, a rotating paddle within the container, significantly increased efficiency. This relatively simple addition markedly reduced the physical exertion required, a considerable improvement for those undertaking this daily chore. The dasher's design, often intricately carved or cast, also became a point of artistic expression, transforming a functional object into a piece of folk art. Variations in dasher design, from simple paddles to more complex multi-bladed structures, are indicative of the ongoing quest for optimal butter production. The materials used also shifted over time, reflecting advancements in metalworking and manufacturing.

Metal churns, particularly those made of tin or painted steel, became increasingly popular during the industrial revolution. Their durability and ease of cleaning were significant advantages over their wooden counterparts. However, even with these advances, the fundamental principle remained the same: the controlled agitation of cream. The introduction of the crank, attached to the dasher, allowed for a more consistent and less strenuous churning process. The crank, a relatively simple device, represents a significant leap in design, allowing for more control and less physical strain. The addition of this element, along with the adoption of stronger and more durable materials, transformed the butter churn from a laborious chore to a relatively straightforward process.

The evolution of the butter churn is also tied to the changing cultural landscape. In many communities, the butter churn became more than just a kitchen utensil; it was a symbol of domesticity, a representation of self-sufficiency, and sometimes even a social hub where neighbours would gather to assist with the butter-making process. Its presence n family homes speaks to its deep integration into daily life, mark-ng a period where food preservation and preparation were integral to ousehold management. The size and style of the churn often varied lepending on the scale of production; large communal churns were em-loyed in areas with extensive dairy operations, while smaller family-ized churns were found in smaller farms or households.

)etailed examination of antique butter churns reveals a rich tapestry f techniques, including decorative elements and stylistic choices hat reflect regional traditions and evolving manufacturing practices. ome churns feature intricate carvings, reflecting the craftsmanship of heir creators and the cultural context in which they were produced.)thers might showcase painted designs, or the use of specialized met-ls, adding to the diversity of this seemingly simple object. These deco-ative elements often include floral patterns, geometric shapes, or even lepictions of rural scenes, reflecting the local artistic styles and tastes. he careful study of these details not only enhances our understanding f the objects themselves but also provides invaluable insights into the ives and cultures of the people who created and used them. These ele-nents tell us as much, if not more, about the lives of their owners than he functionality of the churn itself. The careful study of these artifacts rovides a window into the lives and ingenuity of generations past.

he materials used in butter churn construction also provide valuable lues about technological advancements. The transition from wood to metal, for instance, reflects the increasing availability and affordabil ity of metalworking techniques. The types of wood used—whether i was readily available softwood or a more durable hardwood—reflec both local resources and the desired lifespan of the churn. Simi larly, the use of specific metals—like tin, cast iron, or even paintec steel—provides evidence of advances in metallurgical practices anc manufacturing techniques. Even seemingly insignificant details, lik the type of fasteners used, can offer insights into the evolving tools anc techniques employed in the construction of these objects. This combina tion of artistic expression, technical innovation and social context cre ates a compelling narrative reflected in the evolution of the seemingly simple butter churn.

5 Chapter 5: Navigation & Surveying

The astrolabe, a marvel of ancient technology, deserves a place of honor in any collection. Its intricate brass plates, etched with celestial coordinates, allowed sailors and astronomers to determine latitude, time, and even predict the positions of stars and planets. Holding one, you can almost feel the weight of centuries of exploration and discovery. Its elegance is matched only by its practicality, a testament to the ingenuity of its creators. Different cultures refined the astrolabe's design over centuries, each iteration reflecting a deeper understanding of celestial mechanics. Studying these variations unveils a fascinating glimpse into the historical development of navigation and astronomy. The sheer precision involved in crafting these instruments, with their moving parts and delicate engravings, is truly awe-inspiring.

Early compasses, far simpler in design than the astrolabe, were nonetheless revolutionary. Imagine the trepidation, yet the immense excitement, of early sailors relying on a magnetized needle to guide them across vast, uncharted oceans. These early compasses, often crudely fashioned, represented a giant leap forward in navigation. They were not merely practical tools; they unlocked exploration on a scale previously unimaginable. From primitive floating needles in bowls of water to more sophisticated designs with magnetic needles pivoted on a point, their evolution tells a story of relentless human curiosity and the brave pursuit of the unknown. The impact on global trade and cultural exchange is undeniable.

The theodolite, a sophisticated instrument used in surveying, represents a remarkable achievement in precision engineering. Unlike the simpler tools of earlier eras, the theodolite allowed for the incredibly precise measurement of angles, crucial for creating accurate maps and plans. Its development marked a significant advancement in land surveying, facilitating the construction of complex infrastructure like canals, roads, and cities. The intricate mechanisms, involving finely tuned gears and calibrated scales, demanded exceptional craftsmanship. Examine a well-preserved theodolite and you will appreciate the meticulous work of its creators, who built it for accuracy and longevity. These instruments enabled large-scale projects that previously were considered impossible, reshaping landscapes and civilizations. The early theodolites are silent witnesses to the ambitious projects that transformed the world.

While seemingly simple, the construction of these navigational and surveying tools involved significant skill and expertise. The materials themselves—brass, iron, wood, and sometimes precious stones—were carefully selected for their durability and resistance to corrosion. The intricate machining and engraving were achieved using primitive yet effective techniques, passed down through generations of skilled artisans. These tools were not mass-produced; each one was meticulously crafted, reflecting the pride and dedication of its maker. The tools themselves speak of a time when craftsmanship was valued above all else and are exquisite examples of functionality and artistry combined. Understanding the materials and techniques involved adds a deeper appreciation for their ingenuity.

Consider the challenges faced by early navigators and surveyors. They relied on these tools not only to find their way but also to map and understand the world around them. Their explorations, often fraught with danger and uncertainty, expanded our understanding of geography and opened up new trade routes. The instruments they used were their trusted companions, guiding them across vast oceans and treacherous terrains. The legacy of these individuals and their tools is profound and enduring, shaping our world in ways we are only beginning to fully appreciate. The history of navigation and surveying is inseparable from the development of the ingenious tools described here. Their continued study provides invaluable insights into the human spirit of discovery and our ongoing quest to understand the world.

5.1 Astrolabes & Globes

The astrolabe, a marvel of ancient engineering, deserves a place of honor in any collection of antique instruments. Its intricate brass or bronze discs, etched with celestial coordinates, allowed navigators and astronomers alike to determine the time, latitude, and the positions of stars and planets. More than a simple tool, the astrolabe represents a sophisticated understanding of spherical trigonometry and the mechanics of the heavens. Its portability and adaptability made it indispensable for centuries, aiding explorers in charting unknown waters and assisting scholars in their astronomical observations. Mastering its use required considerable skill and knowledge, making proficient astrolabe users highly valued individuals.

A key aspect of appreciating astrolabes lies in understanding their construction. Each component—the mater, rete, tympans, and rule—plays a specific role in the instrument's function. The careful craftsmanship evident in surviving examples speaks volumes about the precision and dedication of the artisans who created them. The intricate engravings, often featuring astrological symbols or astronomical details, add to their aesthetic appeal, reflecting the intertwined nature of astronomy and astrology during many periods of their use. Examining the mateials used—the type of metal, the style of engraving, the presence of any ewels or inlays—can often provide clues to its origin and the time peiod of its manufacture.

lobes, on the other hand, offered a different perspective on the cosnos. These spherical representations, whether celestial or terrestrial, rovided a three-dimensional visualization of the known world or the eavens. Early globes, often crafted from wood and painstakingly and-painted, were both scientifically informative and artistically imressive. They provided a compelling visual aid for understanding geographic locations, relative distances, and the overall shape and strucure of the Earth. Consider the challenges involved in creating these bjects: accurately mapping the continents, representing the oceans, nd depicting the constellations—all on a curved surface. The accuracy nd detail present in many surviving globes stand as testaments to the napmakers' skill and dedication.

he collecting of astrolabes and globes presents unique challenges and ewards. Authenticity is paramount, requiring careful scrutiny of the onstruction techniques, materials, and any markings present. Reearching the historical context is essential, determining the instrunent's maker, date of manufacture, and any significant historical asociations. The condition of the instrument is also a crucial factor, imacting both its value and its scientific significance. Restoration, when eeded, must be undertaken with extreme care, respecting the historical integrity of the object.

Beyond their scientific and historical value, astrolabes and globes possess an undeniable aesthetic appeal. Their intricate designs, often combining scientific precision with artistic flair, make them captivating objects of study. They serve as a tangible link to past generations of scholars, explorers, and artisans, reminding us of humanity's persistent curiosity about the universe and our place within it. The detailed ornamentation, the carefully chosen materials, and the inherent beauty of the spherical form combine to create pieces that are both functional and visually stunning. Displaying and preserving such instruments demands both respect for their historical significance and an appreciation for their exquisite craftsmanship. They are not merely tools; they are works of art that merit careful study and preservation for generations to come. Collectors should carefully consider proper storage and handling techniques to protect these delicate and historically significant objects, ensuring that they remain a testament to the ingenuity of past civilizations for many years. The acquisition of either of these historical artifacts should be considered with careful research and planning to ensure both the authenticity and the proper care and handling of the specific object are considered.

5.2 Theodolite History

The theodolite, a seemingly simple instrument, boasts a surprisingly rich history intertwined with advancements in surveying and engineering. Its origins are not pinpointed to a single inventor or date, but rather a gradual evolution spanning centuries. Early forms, rudimentary in comparison to their modern counterparts, relied on basic geometric principles and readily available materials. Imagine, if you will, ancient civilizations utilizing rudimentary levels and plumb bobs to establish verticality and horizontal alignment – the foundational elements upon which more sophisticated instruments like the theodolite would later build. These early devices, though lacking the precision of their successors, played a vital role in construction projects, irrigation systems, and the mapping of land. The need to accurately measure angles, critical for surveying and construction, drove innovation.

Improvements came slowly, with each iteration refining the accuracy and ease of use. The incorporation of graduated circles, allowing for precise angular measurements, marked a significant step forward. Early versions might have utilized simple sighting devices, but advancements in optics later led to the integration of telescopes, greatly enhancing the precision and range of these instruments. This transition is mirrored in other instruments, like telescopes themselves, illustrating a broad trend of improving accuracy via optical enhancements.

Think of the painstaking craftsmanship involved—each carefully calibrated degree marking, each precisely aligned component, a testament to the skill and dedication of the artisans and engineers who created these tools.

The development of the theodolite was intrinsically linked to the expanding demands of increasingly complex engineering projects. The construction of grand cathedrals, sprawling fortifications, and intricate canal systems all necessitated precise measurements. The growing importance of land surveying, particularly for property rights and taxation, also spurred further innovation. Governments and institutions invested in improving surveying techniques, leading to a rise in the production of increasingly sophisticated instruments. This highlights the symbiotic relationship between technological advancement and societal needs. Demand drove supply, and the subsequent supply further fueled technological breakthroughs.

The 18th and 19th centuries witnessed significant leaps in the theodolite's design. The incorporation of finely crafted gears, screws, and improved optical lenses allowed for unparalleled accuracy. These improvements were not merely incremental; they represent a paradigm shift in the capabilities of surveying instruments. Manufacturers began specializing in the creation of these tools, leading to a wider availability and standardization of designs. This period also saw the emergence of specialized theodolites designed for specific tasks, such as mining surveying or astronomical observations.

By the early 20th century, the theodolite had largely reached its mature form. While modern electronics have since further enhanced precision and automation, the fundamental principles of the instrument remained largely unchanged. The legacy of the theodolite extends beyond its practical applications; it stands as a symbol of human ingenuity, perseverance, and the unwavering pursuit of precision. The instrument's evolution reflects our continuous striving to measure and understand our world more accurately. Consider its importance; from humble beginnings as a simple tool for leveling, the theodolite grew into the complex optical instrument that it is today – and this growth closely mirrors the developments of the surveying profession. The theodolite, therefore, serves as a tangible embodiment of the progress made in surveying and engineering.

The study of antique theodolites offers a unique window into the past. Examining their construction reveals much about the materials and techniques available to past generations. Variations in design reflect the differing needs and priorities of different cultures and time periods. Collecting and preserving these instruments not only honors the craftsmanship and ingenuity of their creators, but also contributes to our understanding of the historical development of surveying and engineering. Every preserved example speaks to the dedication of these early scientists and artisans. They stand as testament to their dedication to creating instruments of precise measurement.

The appreciation for antique theodolites is not confined to historians and collectors. These instruments offer a tangible link to the history of science and technology, reminding us of the incremental progress that has led to the sophisticated tools we use today. The elegance of their design, the precision of their mechanics, and the historical context in which they were created all contribute to their enduring appeal. It is a reminder of how much the past affects our present.

Beyond their historical significance, antique theodolites hold a certain aesthetic charm. Their intricate mechanics and elegant designs, often crafted from brass or other durable metals, are pleasing to the eye. Many examples survive even today, functioning as beautiful objects of art and design. The enduring appeal of such instruments rests in both functionality and appearance. This balance of art and function speaks to the artistry involved in their creation and use.

Therefore, the exploration of the theodolite's history provides more than just a technical overview of surveying instruments. It offers a compelling narrative of human innovation, reflecting our constant strive for precision and our unwavering curiosity about the world around us. The journey from rudimentary leveling tools to the sophisticated instruments of today is a testament to human ingenuity and perseverance. And, the continued preservation and study of these artifacts ensures that this rich legacy will continue to inspire and inform uture generations of engineers, surveyors, and enthusiasts.

5.3 Early Compasses

he earliest compasses, far from the sleek, precise instruments we know oday, were surprisingly simple devices. Their origins are shrouded n some mystery, with evidence suggesting independent invention in everal parts of the world. One of the earliest forms utilized a magne-ized needle, often a sliver of lodestone, carefully floated on water in bowl. This ingenious method allowed the needle to align itself with he Earth's magnetic field, providing a rudimentary directional indica-or. The simplicity belied its profound impact, marking a crucial step in avigation, opening previously unimaginable possibilities for explo-ation and trade.

hese early floating needle compasses, though effective, suffered from imitations. The slightest breeze could disrupt the delicate balance of he needle, rendering the reading inaccurate, or even useless. Further-nore, the absence of a fixed scale prevented precise angular measure-nents, making navigation reliant on experience and intuition rather han a finely calibrated instrument. Such imperfections highlight the ngoing process of refinement and innovation in instrument making hroughout history. Improvements, while gradual, significantly en-anced the reliability and precision of these crucial navigational tools.

The design of these early compasses evolved, moving from simple bowl of water to more sophisticated containers. These containers were of ten made of brass or other durable materials, offering better protectio from the elements and potentially reducing the effect of wind. Som early examples included a small, shallow, magnetized needle place atop a pivot point on a circular card, reducing friction and allowing fo slightly more stable readings. These pivoting-needle compasses were noticeable step toward the more advanced, dry-compass designs tha would follow. Their inherent stability proved a significant improve ment over their floating counterparts, making them more reliable eve in challenging conditions at sea.

Dry compasses, marking a substantial leap in design, effectively ad dressed the limitations of their predecessors. They eliminated the de pendence on water, using a carefully balanced needle that spun freel on a pivot point within a protective casing. This innovation not onl enhanced the instrument's stability against the wind and waves bu also allowed for greater precision. Early dry compasses, often encase in simple brass or wooden boxes, became increasingly common amon navigators. This crucial development drastically improved accurac and reliability of navigational data and paved the way for further ad vancements.

Further enhancements to early compass technology came in the form of improved needle design. While initially employing naturally mag netized lodestone, later compasses utilized needles made of hardened steel, artificially magnetized for greater strength and consistency. This crucial step yielded significantly more accurate readings, reducing errors and improving the overall performance of the instruments. The process of magnetization itself evolved, with various methods employed to increase the strength and durability of the magnetic field within the compass needles.

The addition of a compass rose, a circular card marked with directional points (cardinal directions and intermediate points), represented a substantial advance. This innovation enhanced the precision and ease of use of the compass. The compass rose enabled navigators to not only determine direction but to make more accurate measurements of angles and bearings. This ability facilitated more precise navigation, opening possibilities for more ambitious voyages of exploration, significantly influencing the course of global maritime history.

The evolution of early compasses was a gradual but significant process. From simple floating needles to more sophisticated dry-card compasses with rose markings, each iteration reflected the relentless pursuit of improvement by skilled artisans and navigators alike. The improvements weren't merely incremental adjustments but rather fundamental changes that enhanced accuracy, reliability, and ease of use, ultimately transforming navigation and exploration, and leaving an indelible mark on history. These early compasses, despite their apparent simplicity, represent a remarkable testament to human ingenuity and the enduring quest for knowledge and exploration. Their impact extended far beyond navigation, influencing cartography, trade, and our understanding of the world. The seemingly small step towards understanding magnetic fields had enormous consequences.

6 Chapter 6: Printing & Writing

The invention of the printing press irrevocably altered the course of human history. Before Gutenberg, the painstaking process of hand-copying manuscripts limited the dissemination of knowledge to a privileged few. Imagine the sheer effort required to duplicate a single book—a labor-intensive process involving months, even years of dedicated work. Now, consider the impact of mechanical reproduction, capable of generating hundreds, thousands, even millions of identical copies with remarkable speed. This democratization of knowledge laid the foundation for the Renaissance, the Reformation, and the Scientific Revolution, each propelled by the ready availability of printed materials. The ramifications are still felt today.

Early printing techniques relied on woodblocks, carved with painstaking detail to produce a single page. Master craftsmen, wielding chisels and gouges with astonishing skill, would meticulously carve away sections of the wood, leaving behind the raised image. Ink was then applied to the surface, transferred to paper, resulting in a reversed image. The process was slow, the output limited, yet it represented a monumental leap forward from the purely handwritten manuscript. The sheer artistry and physical endurance required are worthy of our greatest admiration. It required enormous skill and patience to create such illustrations.

Movable type, a seemingly simple yet revolutionary innovation, exponentially increased printing efficiency. Individual letters, crafted from metal, could be arranged and rearranged to compose lines of text. This allowed for the creation of multiple pages from a single set of type, vastly reducing both time and cost. Think of the logistical challenges: the need for precise casting, consistent sizing, and efficient storage and organization of thousands of individual pieces, all demanding extraordinary organizational and technical skills. The organization and planning required was immense, even for a small project.

Beyond the press itself, the production of books required a complex interplay of tools and techniques. Quills, fashioned from bird feathers, were the primary writing implements. These delicate tools required skillful manipulation, each writer possessing their own technique. The inkwells, often crafted from materials such as horn or ceramic, held the viscous ink, a carefully prepared mixture of pigments, binders, and water. The quality of the ink, the consistency of the nib, and the steady and of the scribe all contributed to the final product. The resulting ext, often meticulously illuminated with hand-painted illustrations, vas a testament to the combined efforts of numerous craftspeople.

Voodcut and engraving techniques further enhanced the aesthetic ap-eal and informational content of printed materials. Woodcuts, em-loying a relief printing process similar to that of the woodblock print-ng press, allowed for the integration of illustrative images within the ext. Engraving, on the other hand, utilized a completely different pro-ess, involving precise incisions into a metal plate, usually copper. This llowed for much finer detail and greater precision, producing incred-bly sharp and intricate images. The tools themselves—burins, gravers, nd scrapers—were specialized instruments requiring years of dedi-ated practice to master. The creation of a single well-done engraving ould take days, even weeks to complete.

'inally, the production of paper itself was a crucial element of the rinting process. Early papermaking methods involved numerous teps, from the preparation of fibrous material such as linen rags to he intricate process of pulping, sheet formation, and pressing. Water-narks, subtle yet distinctive patterns incorporated into the paper dur-ng the manufacturing process, served as both aesthetic and practical eatures. They helped identify the paper's origin and quality. Consider he environmental and economic factors involved: the collection of raw naterials, the energy needed for the manufacturing process, and the distribution network needed to bring the finished product to marke The humble sheet of paper represented the culmination of a comple manufacturing process, involving several stages that demanded bot physical labor and technical skill. The production of paper, printin and binding was an intricate chain of processes. All these steps forme a complex manufacturing process requiring collaboration among var ious specialists. All these were necessary for the creation of printe books.

6.1 Early Printing Presses

The earliest printing methods, predating the Gutenberg press, relie heavily on woodblocks. Imagine the painstaking process: artists metic ulously carving entire pages into wood, a labor-intensive undertak ing demanding immense skill and patience. Each block, representin a single page, was inked and then pressed against paper, yielding single print. This process, while effective for producing images an text, was inefficient for large-scale printing, limiting the dissemina tion of knowledge and information. The creation of multiple identica copies of a single work was incredibly time-consuming and expensive This woodcut printing, however, laid the foundation for later innova tions, showcasing the immense creativity and ingenuity of early print ers even with their basic technologies. It also offers a rich insight int the value placed on written and illustrated materials, a clear demonstration of the importance of both artistic and practical skills in early printmaking.

These limitations spurred innovation. The concept of movable type, where individual characters could be arranged and rearranged, emerged gradually across various cultures. Consider the Chinese, who experimented with movable type using clay and wood as early as the 11th century. While these materials presented challenges, they represent a significant conceptual leap—the realization that assembling characters individually offered significantly increased efficiency. However, these early attempts often suffered from the fragility of the material used. The sheer number of characters required in many languages, for example, in China, contributed to inherent complexities and practical limitations. The durability of the types used greatly affected the speed and efficiency of the process.

Johannes Gutenberg's printing press in the mid-15th century, often hailed as a revolutionary advancement, was not a sudden, isolated invention. It built upon the existing knowledge and techniques of metalworking, typography, and printing that had existed in various forms throughout history. Gutenberg's innovation lay not just in the individual components—but in their ingenious combination. His press utilized a screw-type mechanism for applying even pressure to the inked type, effectively transferring the characters to the page. The process was still incredibly demanding, requiring immense physical effort, and the high cost of creating metal type also meant that it remained restricted to wealthy patrons and institutions. Yet it was this system that finally allowed for mass production of books, making knowledge more widely accessible than ever before. The ability to create multiple copies quickly facilitated the dissemination of ideas and information, forever changing the landscape of communication and scholarship.

The materials used in early printing presses were also crucial to their success. Wood was readily available, and served as the foundation for many early presses. Metal, however, proved superior in terms of durability and precision. The development of specialized alloys and the advancement of metalworking techniques were instrumental in creating types that could withstand repeated use and maintain sharpness. The quality of the metal directly impacted the clarity and longevity of the printed text, showcasing the significant role of materials science in early printing technologies. This is evident in the evolution of the typefaces themselves, reflecting both artistic sensibilities and the constraints of the materials available. This constant interplay between the available resources and innovative ingenuity led to the refinement and widespread adoption of printing methods, changing our world forever.

Beyond the press itself, the creation of high-quality paper was vital. Early papermaking techniques involved a lengthy and arduous process, requiring skilled labor and specific materials. The quality of the paper directly affected the clarity of the print, and the availability of paper significantly impacted the spread of printing technology. The consistent supply of durable, affordable paper was a fundamental requirement for the widespread success of printing presses. This is a point that is easily overlooked when focusing on the press alone. The development of papermaking—a seemingly separate industry—was intrinsically linked to the advancement of printing.

The impact of early printing presses is immeasurable. The increased accessibility of books fueled the Renaissance, the Reformation, and the Scientific Revolution. Ideas could travel faster and reach wider audiences, fostering intellectual exchange and debate on an unprecedented scale. Scholars could access primary sources and build upon each other's work, leading to significant advancements in various fields of knowledge. The ability to mass-produce written materials fundamentally changed societies across the globe. The accessibility to literature and information fueled the social and intellectual changes that defined those eras. This increased access to information had, and continues to have, an enduring legacy. The early printing press was more than a tool; it was a catalyst for transformative societal change.

6.2 Quills & Inkwells

The humble quill, a seemingly simple instrument, holds a significant place in the history of written communication. Its delicate point, fashioned from a bird's feather, allowed for a level of precision and nuance unavailable with earlier writing tools. Consider the painstaking care taken in preparing these quills: the selection of the right feather, the precise cutting and shaping, the delicate slitting of the tip to create the perfect writing surface. Each quill was a unique tool, its characteristics influencing the writer's style and the very texture of the words on the page. This individualized nature contributes to the allure of antique quills for collectors, each representing a tiny piece of history.

Inkwells, the companions of quills, equally deserve our attention. These vessels, ranging from simple ceramic pots to elaborately crafted silver or pewter containers, served as repositories for the vital writing fluid. The design of the inkwell wasn't merely functional. Many were miniature works of art, demonstrating the craftsmanship and artistic sensibilities of their creators. From simple, utilitarian designs to ornate pieces embellished with intricate carvings or precious metals, these inkwells reflected the social standing of their owners and the value placed on the written word. The shapes, materials, and decorations offer a fascinating glimpse into the aesthetic preferences and social hierarchies of past eras.

he interaction between quill and inkwell was more than just a practical pairing; it represented a ritual. The careful dipping of the quill nto the ink, the precise control needed to avoid blotting, the rhythmic cratching of the feather against the parchment—these actions were inegral to the writing process. It was a slow, deliberate act, demanding ocus and patience, qualities that imbued the written word with a cerain gravitas absent from the speed of modern typing. The very act of vriting with a quill and inkwell was a meditation, fostering a connecion between the writer and the message being created. The physicality f the act also enhanced memory retention, studies have shown.

he materials used in the creation of quills and inkwells varied greatly lepending on geographic location and available resources. Quills, prinarily derived from goose, swan, or turkey feathers, were carefully elected and prepared for writing. The ink itself was a complex mixure, its formulation evolving over centuries. Early inks were often nade from natural ingredients such as soot, gallnuts, or various plant xtracts, resulting in a broad range of colors and consistencies. The ecipes themselves, sometimes passed down through generations of cribes, were closely guarded secrets, a testament to the importance of uality writing materials.

Beyond their practical function, antique quills and inkwells offer valuble insights into the socio-economic context of their time. The materils used, the level of craftsmanship, and even the decorative elements can reveal clues about the social status and profession of the owner. For example, a finely crafted silver inkwell might have belonged to a wealthy merchant or a prominent scholar, while a more humble earthenware inkwell may have belonged to a student or a clerk. The condition of these pieces, as well as any accompanying marks or inscriptions, can also offer valuable information about their provenance and usage. Careful examination provides a window into the lives and professions of those who used them.

Collecting antique quills and inkwells is more than just accumulating artifacts; it's about preserving a tangible link to the past. These objects represent not just tools of writing, but also emblems of a time when the written word held a different weight and significance. The physicality of the writing process, the attention to detail required, and the intrinsic connection between the writer and the tool all contributed to a more deliberate and thoughtful approach to communication. Understanding the history and craftsmanship of these items allows collectors to connect with previous generations of writers and artists, offering unique insights into history and human ingenuity. The careful study of these implements also presents an opportunity to appreciate the beauty of simplicity and the enduring value of timeless design. Their quiet elegance testifies to the enduring power of handcrafted tools. The delicate balance between functionality and artistry present in many antique quills and inkwells speaks to a skillset and patience that are increas ingly rare in our fast-paced, technologically driven society. Their rarity today only enhances their charm and historical value.

6.3 Woodcut Tools

The creation of a woodcut, a seemingly simple process, relies on a surprising array of specialized tools. These aren't merely rudimentary carving implements; rather, they represent a sophisticated toolkit developed over centuries to achieve precise detail and nuanced expression in the artistic medium. Understanding these tools allows a deeper appreciation for the skill and artistry involved in producing fine woodcuts.

The most fundamental tool is, of course, the carving knife. However, this isn't just any knife; woodcutters use a variety of knives, each designed for a specific purpose. Some have broader blades for removing large areas of wood, while others possess narrower, more finely shaped blades ideal for intricate details. The handles themselves often exhibit ergonomic designs, crafted for comfort and control during hours of meticulous work. The selection of wood also influences the knife's effectiveness, with harder woods demanding a sturdier, more robust blade than softer ones.

Beyond the knives, specialized gouges play a crucial role. These tools, with their curved blades, allow artists to excavate recesses and create areas of varying depth within the woodblock. Different gouges possess varying curves and widths, providing the artist with a range of options to achieve the desired effect. The sharpness of these tools is paramount; a dull gouge can lead to ragged edges and imprecise lines, compromising the final image's quality. Sharpening and maintaining these gouges is a skill learned through years of practice. Proper honing techniques and the use of appropriate sharpening stones are essential for maintaining their efficiency.

Mallet and hammer are frequently misunderstood tools. A simple mallet, while seemingly unassuming, requires significant finesse. The correct weight and type of mallet, along with the precise application of force, are key factors in effective woodcutting. Too much force, and the wood will shatter; too little, and the cutting process will be slow and inefficient. The hammer, generally used with chisels, aids in creating more robust cuts, essential for removing larger areas of wood quickly without damaging the underlying image. The choice of mallet and hammer head material matters, too, impacting impact and durability. Wooden mallets absorb shock better than metal ones, reducing the risk of damaging the woodblock.

Many may overlook the importance of specialized rulers and gauges. Accurate measurements are critical in creating woodcuts that align properly and produce an even, clear print. These aren't standard drafting tools, but instead often feature particular designs, suitable for the unique constraints of woodcut creation. The incorporation of these tools ensures precision and alignment across the entire block, particularly crucial in multi-block prints where precise registration is essential for aligning different sections of the print. Precision is of the utmost importance when creating intricate woodcuts with multiple blocks.

Beyond the cutting and measuring instruments, one finds a surprising array of finishing tools. These tools, which may seem incidental, are critical in refining the print's final presentation. Burnishers, for instance, smooth the edges of the cut lines, enhancing the print's clarity and sharpness. These are used to compress the wood fibers and create a level surface. Various types of brushes are used for cleaning the block and applying inks. Even seemingly simple tools, like scrapers and small knives, are employed for cleaning debris from the block's surface and for delicately removing minor imperfections or burrs.

Finally, no discussion of woodcut tools is complete without mentioning the importance of proper storage and maintenance. Woodcutting tools, especially those with delicate blades, require careful handling and storage to maintain their sharpness and longevity. Protective sheaths or cases safeguard the blades from damage, preserving their integrity and ensuring the long-term usability of the implements. Regular cleaning and oiling prevent rust and extend the tools' lifespan, a critical consideration given the expense and difficulty in obtaining some of these specialized implements. A keen woodcutter values their tools not merely as implements, but as extensions of their own artistic skill and vision. The craft demands respect for these tools, reflecting the respect for the art form itself.

6.4 Engraving Techniques

Engraving, a process seemingly simple in its concept—incising a design into a surface—reveals astonishing complexity when one delves into its diverse techniques and historical evolution. From the earliest cave paintings etched onto stone to the intricate details of a finely crafted copperplate, the artistry and skill involved are undeniable. The tools themselves, often simple in appearance, represent a sophisticated understanding of materials and their interaction. Consider the burin, for example, a seemingly unassuming steel rod with a sharply angled point, yet capable of producing lines of remarkable delicacy and precision. The subtle variations in its shape and the angle at which it's held dictate the resulting line's character—fine, bold, scratchy, or smooth. Mastering the burin requires years of dedicated practice.

The choice of material greatly impacts the final product's aesthetic and longevity. Wood, a readily available medium, has been used for centuries, yielding woodcuts both bold and detailed. However, the limitations of wood's grain and texture often present challenges, demanding a nuanced understanding of the material's properties. In contrast, netal, specifically copper and steel, offered engravers a vastly different anvas. The smooth, consistent surface allows for incredibly fine lines nd intricate details, resulting in prints of stunning clarity and richess. Preparing these metal plates, however, is a laborious process, inolving meticulous smoothing and polishing to create a perfectly even urface for engraving.

'echniques vary significantly depending on the material and the deired effect. Line engraving, perhaps the most traditional method, inolves meticulously incising lines into the surface using a burin. The epth and spacing of these lines determine the final image's tonal ange and the richness of its details. Stipple engraving, on the other and, employs a very different approach. Instead of continuous lines, he image is created using a multitude of tiny dots, the density and disribution of these dots determining the tonal values. This technique demands an exceptional level of control and precision, resulting in works f unparalleled delicacy. Mezzotint, a unique process, involves creatng a roughened surface on the plate, then selectively smoothing areas o create tonal variations. The result is a velvety, soft texture that is disinctly different from line or stipple engraving. Each technique offers a nique artistic expression, reflecting the engraver's skill and vision.

Beyond the core techniques, various tools enhance the engraver's caabilities. Scratchers, burnishers, and scrapers help refine the lines, often harsh edges, and create subtle tonal variations. These tools,

seemingly minor in their function, allow for a level of control and fi
nesse that transforms the engraving from a mere representation to
work of art. The process is inherently time-consuming, demanding pa
tience, focus, and a deep understanding of both the technical aspect
and the artistic vision. Often, multiple stages of work are required, eac
requiring meticulous care and precision to ensure the final product'
success. The transfer of the image from the engraved plate to the pa
per itself demands a sophisticated understanding of inks and printin
techniques.

Considering the historical context expands our appreciation for the in
genuity of engravers. Early techniques, employing rudimentary tool
and materials, produced remarkable results, demonstrating the re
sourcefulness of early artisans. As technology advanced, so did th
tools and techniques, leading to an explosion of innovation and artisti
expression. The development of specialized tools and processes, reflect
ing the evolving needs and artistic goals, is a testament to human inge
nuity and adaptability. Examining historical examples, from the del
icate engravings adorning ancient jewelry to the intricately detaile
maps produced centuries ago, reveals the profound impact of this ar
form on our understanding of the past. The preservation of these en
graved pieces speaks volumes about their enduring value and artisti
significance. From the humble tools to the magnificent final product
the story of engraving unfolds as a testament to human creativity. Th

evolution of engraving, interwoven with the broader history of art and technology, remains a fascinating subject of study, revealing the constant interplay between artistic expression and technological innovation. Its enduring legacy serves as a powerful reminder of the importance of preserving and celebrating our artistic heritage.

6.5 Papermaking Methods

The earliest methods involved simple hand processes, relying heavily on readily available materials. Think of ancient civilizations utilizing readily accessible plant fibers, like papyrus reeds in Egypt or various tree barks and pulps in other regions. These fibers, after undergoing a meticulous process of soaking, cleaning, and sometimes beating, were then painstakingly layered and pressed, sometimes with the aid of heavy stones or simple wooden presses. The resulting sheets, though crude compared to modern paper, served their purpose admirably, preserving knowledge and facilitating communication in their time. The very nature of these early papers reflected the materials at hand and the ingenuity of their creators.

Later developments saw the introduction of more refined techniques, building upon established processes. The Chinese, for instance, are credited with significant advancements in papermaking, refining techniques that yielded stronger, smoother sheets. This involved improvements in the pulping process, experimenting with different plant fibers and utilizing tools to further break down and refine the material before layering. The use of size, a mixture often comprising animal glue or starch, added further refinement, enhancing the sheet's durability and preventing ink bleed-through. These innovations were vital steps, allowing for the creation of paper suitable for more detailed writing and artistic endeavors.

One particular advancement was the introduction of the "mold and deckle," a revolutionary frame used to shape the paper sheet. This framework, typically constructed from wood and wire mesh, allowed for greater control over the paper's dimensions and uniformity, significantly impacting quality and consistency. The mold and deckle became an indispensable part of the papermaking process, a standard tool that endured for centuries, leading to a higher volume of consistently sized sheets. This was a crucial step towards mass production and accessibility of paper. The standardization resulting from this simple device profoundly influenced the spread of literacy and knowledge across civilizations.

The process itself involved considerable skill and patience. After pulping the fibers, the papermaker would carefully suspend the pulp in water within the mold and deckle. They'd then skillfully agitate the mixture, ensuring an even distribution of fibers. Once a sufficient layer had settled, excess water would be drained, leaving behind a thin layer of intertwined fibers. This layer would then be pressed to remove remaining moisture, ultimately leading to the formation of a paper sheet. This deceptively simple process required considerable dexterity and an intimate understanding of the materials. The quality of the final product hinged heavily on the skill and experience of the papermaker.

Beyond the basic techniques, numerous regional variations and innovative adaptations evolved over time. Different cultures experimented with diverse materials, from linen rags to hemp fibers, creating papers with unique textures, colors, and properties. This process of experimentation and adaptation reflects the dynamic nature of papermaking, a craft continuously evolving to meet changing demands and incorporate newfound knowledge. The resulting variations in paper types, each with its own distinctive qualities, enriched both artistic expression and practical applications. Each method added a nuance to the art of papermaking, producing material that was both utilitarian and aesthetically appealing.

The rise of watermills during the Middle Ages brought about another significant advancement. Previously, pulping fibers was a labor-intensive task, requiring significant manual effort. Watermills provided a mechanical power source, automating the pulping process and greatly increasing production efficiency. This technological shift had a profound impact, enabling papermakers to produce larger quantities of paper, ultimately driving down costs and making it more accessible to a wider audience. The mechanical means increased productivity and reduced the demand for manual labor.

Consequently, the production of paper was no longer exclusively confined to small workshops and monastic environments. The increased efficiency allowed for the establishment of larger-scale paper mills, leading to a significant rise in paper production. This widespread availability of paper, in turn, fueled a growth in literacy, fostering advancements in communication and the wider dissemination of knowledge. The accessibility of paper had a transformative influence on society.

As the demand for paper increased, further refinements and technological advancements followed. The development of improved machinery for pulping, pressing, and drying further enhanced the efficiency and quality of the paper produced. These innovations not only boosted production but also allowed for greater consistency and standardization in paper manufacturing. Mechanization and industrialization transformed the craft of papermaking.

Nevertheless, even with these advancements, handmade paper continued to be valued for its unique qualities and artisanal character. The handmade process preserved an element of individual expression and craftsmanship, creating papers with a distinctive texture and visual appeal. Handmade paper retained its value among artists and calligraphers who appreciated its nuances and superior quality. While mass production created a readily available and cheaper product, the niche or handmade paper persisted.

herefore, understanding the historical progression of papermaking nethods offers a captivating glimpse into human ingenuity and technological advancement. From the rudimentary techniques of early civlizations to the refined processes of later eras, the evolution of papernaking showcases the interplay between human creativity and technological progress. The journey from simple hand processes to complex nachinery is a testament to human innovation and its profound impact on communication and knowledge dissemination. The history of apermaking provides a compelling narrative of this enduring human ndeavor.

7 Chapter 7: Medical Instruments

The development of surgical tools reflects a fascinating interplay between evolving medical understanding and technological innovation. Early surgical implements, often crude by modern standards, were nonetheless crafted with considerable skill. Consider the trephine, a hand-held instrument used for drilling holes in the skull – a procedure, trepanation, practiced for millennia across diverse cultures. While the precise reasons for its use remain debated, its existence demonstrates humanity's enduring attempts to alleviate suffering, even with limited technological means. The evolution of the scalpel, from sharp flints and obsidian blades to the finely crafted steel instruments we know today, exemplifies this persistent drive for refinement. The design and materials continually adapted to minimize trauma, improve precision, and enhance the safety of both the patient and the surgeon. Detailed examination reveals the ingenious methods employed to create strong yet flexible blades, durable handles, and precise mechanisms for delicate operations.

Many early surgical tools were repurposed everyday items, reflecting a practical approach to medical intervention. For instance, simple hooks, clamps, and probes might have been adapted from woodworking or blacksmithing tools. This resourceful adaptation reveals a remarkable level of ingenuity and underscores the scarcity of specialized medical equipment in earlier eras. As medical knowledge expanded, so too did the need for increasingly specialized instruments. The emergence of the forceps, designed to grasp and manipulate tissues during surgery, showcases a move towards more delicate and precise surgical interventions. The intricate design of these instruments, often incorporating multiple joints and precisely calibrated curves, displays a remarkable degree of craftsmanship. Furthermore, the development of specialized retractors, designed to hold back tissues and improve visibility during complex procedures, demonstrates a clear understanding of surgical principles and an appreciation for ergonomic considerations in tool design.

Apothecary equipment offers a distinct, yet equally compelling, perspective on the history of medical technology. The apothecary, a figure central to pre-modern medicine, was responsible for the preparation and dispensing of medicines. Their work involved a range of intricate tools and processes, many of which reflected the alchemical and herbal traditions that underpinned early medicinal practices. Mortars and pestles, ubiquitous in apothecaries, were used to grind herbs, minerals, and other ingredients into fine powders. The design of these simple, yet effective, instruments demonstrates an understanding of material science, ensuring the tools were both durable and able to withstand rigorous use. Detailed examination of surviving examples reveals variations in size, material, and design, reflecting the specific needs and preferences of individual apothecaries and the diverse range of medicinal preparations they undertook.

Balances, another essential piece of apothecary equipment, played a crucial role in accurately measuring ingredients. These delicate instruments, often constructed with meticulously crafted weights and finely balanced beams, highlight the emphasis placed on precision in the preparation of medicines. The slightest inaccuracy could have serious consequences, and the evolution of these balances mirrors the growing demand for precision in the medical field. Detailed analysis of historical balances reveals advancements in design, materials, and sensitivity, reflecting the continuing refinement of measurement techniques in the apothecary's craft. Furthermore, the apothecary's shop frequently included a wide array of containers – bottles, jars, and boxes – often made from glass, ceramic, or wood, each designed to store different medicinal preparations. Their shape, size, and material reflect both functional considerations and aesthetic choices. Careful attention to detail, essential in preventing contamination and preserving the potency of the medicine, influenced the design and construction of each vessel.

The evolution of medical instruments is a powerful testament to human ingenuity and the enduring quest to improve human health. From the earliest crude tools to the highly specialized devices of today, each innovation built upon the knowledge and experiences of generations of practitioners. The study of antique medical implements offers a fascinating glimpse into the history of medicine, revealing not only the technological advancements that have transformed medical practice but also the enduring human desire to alleviate suffering and enhance well-being. The careful preservation and study of these objects allows us to appreciate the ingenuity of those who came before, illuminating the path towards the advanced medical technologies that we benefit from today. These instruments, often overlooked, reveal insights into historical medical practices and technological innovation, providing a rich field of study for anyone interested in the intersection of history and technology. Their intricate designs and materials reflect the evolving understanding of anatomy, physiology, and surgical technique, reminding us of the long and complex journey towards modern medical practices.

ʼ.1 Surgical Tools

he scalpel, perhaps the most iconic surgical tool, underwent a re-
narkable transformation. Early scalpels, often little more than sharp-
ned flints or obsidian blades, were crude yet effective in their sim-
licity. Their use, however, was limited by their lack of precision and
ontrol, leading to potentially significant trauma. The evolution to
ronze and then iron allowed for greater durability and the crafting
f more refined shapes, improving the surgeon's ability to make del-
cate incisions. Later, the incorporation of steel, and eventually spe-
ialized surgical steels, marked a significant leap forward, offering su-
erior sharpness, corrosion resistance, and the ability to maintain a
een edge, a critical factor in minimizing patient trauma and ensur-
ng clean cuts. The development of different blade shapes—from the
ointed blade ideal for delicate work to the broader blades suited for
arger incisions— further refined the scalpel's capabilities.

he forceps, another essential instrument, experienced a parallel evo-
ution. Early forceps were often rudimentary, perhaps fashioned from
imple bone or wood, their design limited by the available materi-
ls and the technologies of the time. Their purpose, to grasp and ma-
ipulate tissue, remained consistent, but their effectiveness increased
ramatically with the advent of metalworking. The development
f more refined designs, incorporating intricate joints and articula-

tions, significantly enhanced their functionality. This allowed sur geons to manipulate tissue with greater precision, minimizing damag to surrounding healthy tissue and making complex procedures mor manageable. Different types of forceps, developed to address specifi needs—such as tissue forceps, artery forceps, and bone forceps—furthe broadened the capabilities of this critical tool. The incorporation of fea tures like serrated jaws, designed to ensure a secure grip on slipper tissue, further improved their usefulness.

Beyond the scalpel and forceps, a wide array of specialized tool emerged over centuries. Retractors, designed to hold back tissue an improve visibility during a surgical procedure, transitioned from ba sic hooks to more sophisticated designs, including self-retaining retrac tors that freed the surgeon's hands. The development of specialized su tures, along with needles designed for their application, revolutionize the closing of wounds. These early sutures, though often made fron simple materials like animal gut or linen thread, were crucial in mini mizing infection and promoting healing. The advancements in materi als science led to the development of stronger, more biocompatible su tures, significantly reducing the risk of complications. The evolutio of surgical saws, from rudimentary hand-held saws to more precis power-driven saws, allowed for the more efficient and less traumati removal of bone. These advancements were not isolated occurrences but rather a continuous interplay between surgical techniques and th development of better tools, reflecting the constant striving to improve surgical outcomes.

The development of anesthesia played a crucial role in the design of surgical tools. Before the advent of general anesthesia, procedures were often short and brutal, requiring tools that could accomplish their task quickly. The introduction of effective anesthesia changed this dynamic. Surgeons now had the time to perform more complex operations with greater precision, requiring more intricate and specialized tools. This fueled innovation, leading to advancements in areas like precision instruments for delicate operations, such as microsurgery. The demand for tools that could handle a longer, more controlled operation created a feedback loop, stimulating the design and development of more advanced equipment.

The history of surgical tools is a testament to human ingenuity and a relentless pursuit of minimizing patient suffering and maximizing successful surgical outcomes. The evolution of these tools, from rudimentary beginnings to the sophisticated instruments of today, mirrors the broader progress of medicine and underscores the crucial link between technological advancements and improved healthcare. Each incremental improvement, no matter how small, reflects the dedication of countless individuals to improve the art and science of surgery. The study of these historical tools offers not only a glimpse into the past but also a valuable lesson in the continuous pursuit of innovation and improvement in medical practice. Careful examination of their design and functionality reveals a fascinating story of problem-solving and adaptation, highlighting the ingenuity of those who shaped the tools that have saved countless lives.

7.2 Apothecary Equipment

Mortars and pestles, the quintessential tools of the apothecary, come in a breathtaking array of sizes, materials, and designs. From simple, rough-hewn stone versions used for centuries in rural settings to the finely crafted porcelain or agate mortars favored by city apothecaries, these tools tell a silent story of meticulous preparation and precise measurements. The pestle, often matching the mortar in material, is used to grind herbs, spices, and minerals into fine powders, a process crucial for creating consistent and effective remedies. The quality of the mortar and pestle itself directly impacts the resulting powder's fineness and consistency, reflecting the apothecary's commitment to quality.

Balances, essential for accurate measurement in pharmaceutical preparations, underwent a fascinating evolution. Early apothecaries relied on simple beam balances, often with weights made from materials like lead or bronze. These balances, while less precise than later models, served their purpose adequately. As time went on, more sophisticated balances were introduced, incorporating features such as fine adjustments and various weight sets to ensure greater accuracy. The development of the torsion balance, with its delicately balanced torsion wire, marked a significant leap forward in precision, minimizing human error and improving the consistency of pharmaceutical concoctions. The intricate details on many antique balances, often adorned with decorative elements, reflect the importance of the profession and the care taken in its practice.

Funnels, seemingly simple tools, played a vital, often overlooked role in the apothecary's work. Used to carefully transfer liquids and powders, funnels came in numerous shapes and sizes, each tailored to a specific task. Those made of glass allowed for better visibility during pouring, reducing the risk of contamination. Funnels of metal, typically brass or tin, were more durable and better suited for transferring coarser materials. The attention to detail in their construction is often subtle, yet significant, indicating the meticulous nature of the work they assisted in. The simple act of pouring, aided by these unassuming tools, was often a key step in the preparation of effective medicines.

Sieves and strainers, often made of fine woven metal or cloth, were integral for separating solids from liquids. The mesh size varied widely, depending on the specific needs of the preparation. Finely woven sieves were used for sifting powdered ingredients, removing lumps and ensuring uniformity. Coarser sieves helped separate larger particles, leaving behind a smooth, consistent material. The material selection for the sieve was also crucial. Metal sieves, such as those made of brass or stainless steel, were durable and easily cleaned, while cloth sieves, often linen or fine mesh, were used for more delicate applications. These seemingly simple sieves played a fundamental role in the process of creating smooth, refined pharmaceutical products.

Finally, storage containers for herbal remedies and prepared medicines warrant special attention. Apothecaries used a wide range of containers, each chosen for its specific properties and suitability to the substance it held. Glass bottles, sometimes sealed with wax stoppers, were often preferred for liquids, their transparency allowing for easy visual inspection of the contents. Earthenware pots, often glazed to prevent leakage, provided sturdy storage for powders and ointments. Wooden boxes, lined with cloth or paper, stored fragile ingredients, protected from damage or moisture. The selection of each container showed the apothecary's understanding of the properties of their materials and their impact on the preservation of valuable ingredients, reflecting a deep appreciation for materials science. The variety and care in selecting such containers underscore the multifaceted nature of the apothecary's trade.

www.ingramcontent.com/pod-product-compliance
Ingram Content Group UK Ltd.
Pitfield, Milton Keynes, MK11 3LW, UK
UKHW020158290926
15471UKWH00005B/933